LIVING IN THE LAND OF DINOSAURS

Ancient Reptiles Come to Life in Augmented Reality

Executive Producer
Steve Grubbs

Science Advisor
Wendy Martin

Curriculum Specialist
Rene Gadelha

Development Team
Cody Birks
Sorine Jille Buendia
Aaron Hagens
Tyler Halterman
Alec Pilola
Nathan Rossi
Tyler Turner

Publisher
VictoryXR

TABLE OF CONTENTS

WHAT IS PANGAEA?

The following illustration shows continental drift: the theory of gradual movement of the continents across Earth's surface through geological time. This image depicts land formation changes from the supercontinent, Pangaea, dating back to 300 million years ago, through the seven continents of today's era.

GEOLOGIC TIME SCALE

EON	ERA	
PHANEROZOIC	CENOZOIC	
	MESOZOIC	
	PALAEOZOIC	LATE
		EARLY

DATE	PERIOD
2.6 MYA	QUATERNARY
23 MYA	NEOGENE
66 MYA	PALAEOGENE
145 MYA	CRETACEOUS
201 MYA	JURASSIC
252 MYA	TRIASSIC
299 MYA	PERMIAN
359 MYA	CARBONIFEROUS
419 MYA	DEVONIAN
444 MYA	SILURIAN
485 MYA	ORDOVICIAN
541 MYA	CAMBRIAN

USER GUIDE

Downloading App Instructions

- Go to the Apple App Store or the Google Play Store and Search "AR Living World of Dinosaurs".
- Download the App.
- Open the App.
- Hold device over the image on the page using the target in the middle of the screen.

For troubleshooting visit:

VictoryXR.com/ARTroubleshoot

Contact us at:

800-670-5716 or info@victoryxr.com

VELOCIRAPTOR

FUN FACT!

They are only about the size of a domesticated turkey!

Bio

They lived in desert like environments. Velociraptors were believed to have been capable of short bursts of speed up to 40mph (60km/hr). A carnivore that was both a predator and a scavenger and were believed to have hunted in packs. Velociraptor Had curved, sickle shaped claws on its hind feet. These were used to slice into prey or as its best method of defense against larger aniamls.

SIZE

Length: 6 Ft
Height: 3 Ft
Weight: 20 - 30 Lb

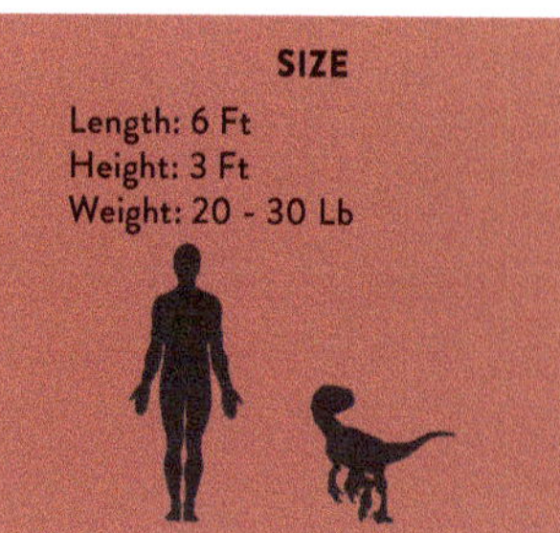

24

NICKNAME

The Swift Thief

GEOLOGICAL TIME

Lived during the late Cretaceous Period in what is now East Asia.

DISCOVERY

First discovered in 1922 in Mongolia.

UNIQUE FEATURES

One of the most bird like dinosaurs ever discovered, many scientists now believe that velociraptors had

25

HOW TO ENGAGE WITH DINOSAURS

National award-winning science teacher Wendy Martin will be guiding you through the Land of Dinosaurs. To engage with the AR marker hover your AR-enabled device over the large image on each page spread. While engaging with the marker you will have the option on the screen to toggle between desktop-sized and real life-sized reptiles! Or you can walk around and view the dinosaurs from all angles.

LEPTOCERATOPS

lep-toe-SAIR-uh-tops

FUN FACT!

It was so much smaller than its relative the Triceratops that Leptoceratops could have walked under the belly of its cousin!

How They Lived

Leptoceratops once roamed grassland and plains areas of what is now North America. They were herbivores who grazed on low-lying plants and grasses.

Their long rear legs indicate that they could walk on all four limbs, as well as rear up on their hind two legs to reach for food or run quickly. The Leptoceratops was quite vulnerable because of its small size and limited armor (body protection).

SIZE

Length: 8 Feet
Height:3 Feet (at the hips)
Weight: 300 Pounds

NICKNAME

Slim-Horned Face

GEOLOGICAL TIME

Late Cretaceous Period

DISCOVERY

North America (1910)

UNIQUE FEATURES

Parrot-like beak and solid neck frill

STYRACOSAURUS

stih-RAK-uh-SAWR-us

FUN FACT!

A large discovery of the fossils of over 100 Styracosauruses were found in Arizona!

How They Lived

The Styracosaurus was discovered along the coastal plains of what is now what is now Europe, Africa, Asia, and North America. These dinosaurs were herbivores that primarily consumed fern and palm plants.

These were social animals that mainly lived in herds. Living with others helped in their defenses, too. Styracosauruses used their horns as a defense against attacking predators.

SIZE

Length: 17 Feet
Height: 8 Feet
Weight: 6,000 Pounds

NICKNAME

Spiked Lizard

GEOLOGICAL TIME

Cretaceous Period

DISCOVERY

North America (1913)

UNIQUE FEATURES

Head frill with
six long spikes

IGUANODON

ig-WAN-oh-don

FUN FACT!

The Iguanodon was the second dinosaur to receive a formal name.

How They Lived

Iguanodons were terrestrial animals found in areas of abundant plant life, primarily in what is now Europe, Africa, Asia, and North America. These were herbivores with teeth similar to modern-day iguanas.

Footprint fossils indicate that Iguanodons traveled in herds, and that they could easily shift between using two legs or all four. Their large thumb spike helped to fend off predators from attack.

SIZE

Length: 30 Feet
Height: 9 Feet
Weight: 7,000 Pounds

NICKNAME

Iguana Tooth

GEOLOGICAL TIME

Early Cretaceous Period

DISCOVERY

Europe (1834)

UNIQUE FEATURES

Teeth resembling modern-day iguanas, but larger

MOSASAURUS

MOH-suh-SAWR-us

FUN FACT!

The Mosasaurus was related to modern-day Komodo dragons.

How They Lived

Mosasauruses were found swimming in the ocean waters between what is now western Europe and North America. These were carnivores believed to be the apex predators of their ecosystem.

They swam and hunted deep in the ocean but had to come to the surface for air. Their double-hinged jaw allowed them to swallow food whole. Mosasauruses dined on giant turtles, marine crocodiles, sharks, plesiosaurs, and seabirds.

SIZE

Length: 55 Feet
Weight: 30,000 Pounds

NICKNAME

Meuse River Lizard

GEOLOGICAL TIME

Late Cretaceous Period

DISCOVERY

Europe, near the Meuse River (1764)

UNIQUE FEATURES

Large eyes, but poor vision; body covered in dark, smooth scales

NODOSAUR

NOHD-o-sor

FUN FACT!

In 2011, a Nodosaur with nearly intact skin, armor and even organs was accidentally discovered by an oil mine worker in Canada.

How They Lived

Although they were terrestrial, Nodosaurs were often found near water in the areas of what is now North America. These were herbivores with small teeth, which suggest they ate soft, less fibrous plant life.

Due to the isolated nature of the fossils found, scientists believe Nodosaurs lived a solitary life. To protect from predators, they likely curled into a ball (like a modern-day hedgehog) to protect their underbelly.

SIZE

Length: 17 Feet
Weight: 4,000 Pounds

NICKNAME

Knobbed Lizard
or
Toothless Lizard

GEOLOGICAL TIME

Early Cretaceous Period

DISCOVERY

North America (1824)

UNIQUE FEATURES

In defense, Nodosaurs hugged the ground, only exposing bony armor

EDMONTONIA

ed-mon-TOE-nee-uh

FUN FACT!

Edmontonia is named for the rock formation in which it was first discovered.

How They Lived

Edmontonia were common in the coastal plains of what is now North America. These were herbivores with small teeth who dined primarily on soft plants.

Male and female Edmontonia may have used their shoulder spikes differently: males fighting for territory and females for shoving contests related to mating. The four large shoulder spikes and bony plates allowed these dinosaurs to defend themselves from any predator's bite.

SIZE

Length: 22 Feet
Weight: 6,000 Pounds

NICKNAME

Lizanton

GEOLOGICAL TIME

Late Cretaceous Period

DISCOVERY

North America (1928)

UNIQUE FEATURES

Had up to 60 rows of teeth on each jaw

PENTACERATOPS

PEN-tah-sair-uh-tops

FUN FACT!

To date it has the largest known skull of any land vertebrate.

How They Lived

Pentaceratops liked to roam in areas of low-lying, abundant plant life of what is now North America. They were herbivores that foraged mosses and ferns close to the ground.

The decorative neck frill of Pentaceratops may have played a role in courtship rituals. It also served as protection from predator bites, as well as being a weapon, used to ram threatening dinosaurs, much like modern-day rhinos.

SIZE

Length: 27 Feet
Height: 14 Feet
Weight: 10,000 Pounds

NICKNAME

Five-Horned Face

GEOLOGICAL TIME

Late Cretaceous Period

DISCOVERY

North America (1921)

UNIQUE FEATURES

Massive head had a parrot-like beak with brow horns

DILOPHOSAURUS

die-LO-fuh-SAWR-us

FUN FACT!

They are NOT believed to have been able to shoot venom as depicted in *Jurassic Park*!

How They Lived

The Dilophosaurus was found living near rivers of what is now North America and eastern Asia. These were carnivores who preyed on small animals and fish.

Believed to have hunted in packs, the Dilophosaurus was a large predator for its time. Despite their size, their jaws indicate that their teeth weren't strong enough to take down predators larger than themselves.

SIZE

Length: 20 Feet
Height: 5 Feet
Weight: 1,000 Pounds

NICKNAME

Two-Crested Lizard

GEOLOGICAL TIME

Early Jurassic Period

DISCOVERY

North America (1942)

UNIQUE FEATURES

Two bony crests
on its head

VELOCIRAPTOR

veh-loss-ih-RAP-tor

FUN FACT!

Velociraptors are only about the size of a domesticated turkey!

How They Lived

The Velociraptor lived in desert-like environments of what is now eastern Asia. They were believed to have been capable of short, rapid bursts of speed up to 40 miles per hour.

Believed to have hunted in packs, Velociraptors were both predators and scavengers. They had curved, sickle-shaped claws on their hind feet that they used to slice into prey or as a defense against larger animals.

SIZE

Length: 6 Feet
Height: 3 Feet
Weight: 25 Pounds

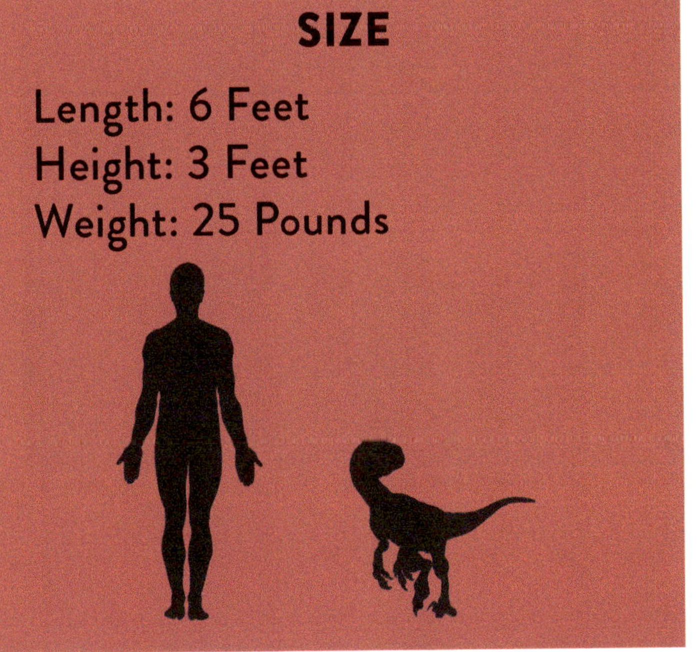

NICKNAME

Swift Thief
or
Rapid Robber

GEOLOGICAL TIME

Late Cretaceous Period

DISCOVERY

Asia (1922)

UNIQUE FEATURES

Had feathers on their bodies and bird-like metabolism

TYRANNOSAURUS

tye-RAN-uh-SAWR-us

FUN FACT!

It was the star of the original *Jurassic Park* movie!

How They Lived

Tyrannosauruses roamed around in forested and swampy areas of what is now North America and Asia. They walked horizontally with their tail erect to help balance their large head.

Scientists argue that the Tyrannosaurus was more of a scavenger than an active hunter, but either way, it possessed large, serrated teeth—some more than a foot long.

SIZE

Length: 40 Feet
Height: 20 Feet
Weight: 14,000 Pounds

NICKNAME

Tyrant Lizard King
or
T-Rex

GEOLOGICAL TIME

Late Cretaceous Period

DISCOVERY

North America (1905)

UNIQUE FEATURES

One of the largest flesh-eating land animals ever

PTERANODON

ter-ANN-uh-don

FUN FACT!

Pteranodon is actually a flying reptile, although it is frequently associated with dinosaurs.

How They Lived

Pteranodon lived in the coastal regions of what is now North America. These were flying reptiles, but they often glided, launching themselves from high objects. On land they walked on all fours instead of hopping around.

They didn't have many natural predators, but occasionally Pteranodons were caught and consumed by large marine reptiles. Their primary food source was fish. Pteranodons fed by scooping up their prey and swallowing it whole since they didn't have teeth.

SIZE

Length: 18 Feet (wingspan)
Height: 6 Feet
Weight: 50 Pounds

NICKNAME

Winged Lizard

GEOLOGICAL TIME

Late Cretaceous Period

DISCOVERY

North America (1870)

UNIQUE FEATURES

Head crest used for balance during flight

ANKYLOSAURUS

ang-KILE-uh-SAWR-us

FUN FACT!

Despite its size, Ankylosaurus is believed to have had a brain the size of a walnut!

How They Lived

Ankylosaurus lived in what is now North America. During this period, they grazed in subtropical forests with plenty of water supply and diverse plant life. They were herbivores with grinding teeth and a horny beak that aided in cutting plant matter when eating.

These dinosaurs were very slow moving but because of their body armor, speed wasn't a necessary defense tactic. The bony armor, spiked head, and clubbed tail of the Ankylosaurus helped defend it from predators.

SIZE

Length: 35 Feet
Height: 8 Feet
Weight: 8,000 Pounds

NICKNAME

Stiff Lizard
or
Fused Lizard

GEOLOGICAL TIME

Late Cretaceous Period

DISCOVERY

North America (1906)

UNIQUE FEATURES

Swinging clubbed tail was lethal to predators

STEGOSAURUS

STEG-uh-SAWR-us

FUN FACT!

Stegosaurus is the only plated dinosaur ever found in western North America.

How They Lived

Stegosaurus lived in what is now North America and Europe. They were herbivores that roamed in the woodland and plains areas, feeding on vegetation. They used their long hind legs to lift off of the ground and eat from tree branches.

The plates on the Stegosaurus were used for thermoregulation (warming them up). Also, its spiked tail and large back spikes were used to ward off predators.

SIZE

Length: 24 Feet
Height: 12 Feet
Weight: 5,000 Pounds

NICKNAME

Plated Lizard

GEOLOGICAL TIME

Late Jurassic Period

DISCOVERY

North America (1877)

UNIQUE FEATURES

Large triangular plates along its back

TRICERATOPS

try-SAIR-uh-tops

FUN FACT!

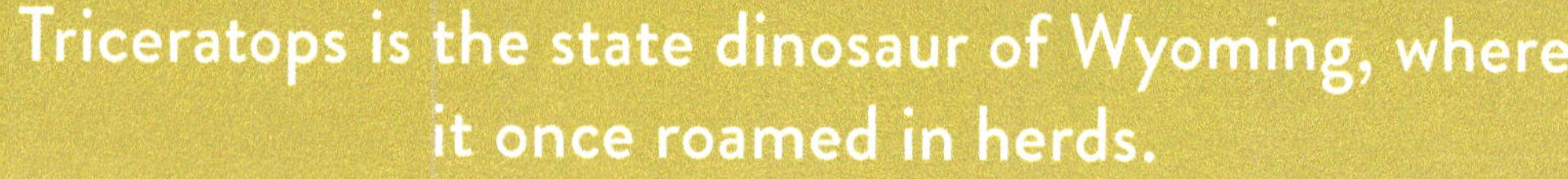
Triceratops is the state dinosaur of Wyoming, where it once roamed in herds.

How They Lived

Triceratops were mostly found in woodland areas of what is now North America. Since many of their fossils were discovered in bone beds of close proximity, it's believed they were herding animals.

An herbivore that ate low growth plants, Triceratops had hundreds of teeth for chewing that often replaced themselves. When threatened by predators, they used their horns to charge their attackers, while their neck frill served as protection from deadly bites.

SIZE

Length: 30 Feet
Height: 10 Feet
Weight: 22,000 Pounds

NICKNAME

Three-Horned Face

GEOLOGICAL TIME

Late Cretaceous Period

DISCOVERY

North America (1887)

UNIQUE FEATURES

Skull and neck frill are almost ⅓ of its total length

BRACHIOSAURUS

BRACK-ee-uh-SAWR-us

FUN FACT!

The large nostrils of this dinosaur were located on top of its head.

How They Lived

Brachiosaurus roamed in the areas of what is now North America and the Mediterrean regions of Europe and Africa. They lived in the semiarid flat plains and ate leaves from tall trees since they were herbivores.

It is believed that the Brachiosaurus held its neck in a high S-shape as opposed to stretching it out completely vertical. It likely defended itself from predators by using its tail or simply smashing attackers underneath its feet.

SIZE

Length: 75 Feet
Height: 40 Feet
Weight: 178,000 Pounds

NICKNAME

Arm Lizard

GEOLOGICAL TIME

Late Jurassic Period

DISCOVERY

North America (1903)

UNIQUE FEATURES

The two front legs were much longer than the back two

PARASAUROLOPHUS

par-ah-SAWR-OL-uh-fus

FUN FACT!

It foraged on all four legs but ran on only two.

How They Lived

Parasaurolophuses were most commonly found in forested areas of what is now North America. They were herbivores that grazed on plant life while traveling in herds.

Their good vision and hearing made them constantly aware of their surroundings. Using their senses to their advantage, they could detect predators quickly. When they had to run from attacking animals, they used their two hind limbs for faster getaways.

SIZE

Length: 35 Feet
Height: 15 Feet
Weight: 6,000 Pounds

NICKNAME

Near-Crested Lizard

GEOLOGICAL TIME

Late Cretaceous Period

DISCOVERY

North America (1922)

UNIQUE FEATURES

Crest used for species and mate recognition, temperature regulation, and sound resonance

SPINOSAURUS

SPY-nuh-SAWR-us

FUN FACT!

Scientists believe Spinosaurus was larger than T-Rex based on fossil measurements.

How They Lived

Spinosaurus lived in the areas of what is now Africa. These animals were carnivores that fed on fish that lived in the dinosaur's river system.

Its nostril openings, crocodilian-like teeth, webbed feet, and dense bones implied it lived mainly in water. Spinosaurus didn't have many predators, but when it needed to defend itself, its large sail was used for that purpose.

SIZE

Length: 45 Feet
Height: 24 Feet
Weight: 15,000 Pounds

NICKNAME

Spine Lizard

GEOLOGICAL TIME

Late Cretaceous Period

DISCOVERY

Africa (1912)

UNIQUE FEATURES

Dorsal fin used for temperature regulation and basking

PLESIOSAURUS

PLEH-see-oo-SAWR-us

FUN FACT!

Cryptozoologists who study mythical creatures believe that the Loch Ness Monster may be a modern-day Plesiosaur!

How They Lived

Plesiosaurus lived in the area of what is now Europe. Its nicknamed "Near-To" because, although it is a prehistoric reptile, it isn't considered a dinosaur since it didn't live on land.

Considered the top apex predator in the ocean of its time, Plesiosaurus was a carnivore that swallowed stones to help it digest larger prey or hard-shelled mollusks. It also swam close to the water's surface using four flippers and its long neck to grab fish and cephalopods.

SIZE

Length: 50 Feet
Weight: 1,000 Pounds

NICKNAME

Near-To Lizard

GEOLOGICAL TIME

Early Jurassic Period

DISCOVERY

Europe (1821)

UNIQUE FEATURES

Long, broad flat body with short tail

ALLOSAURUS

AL-uh-SAWR-us

FUN FACT!

Allosauruses were speed hunters that could run up to 20 miles per hour.

How They Lived

Allosauruses roamed the areas of what is now North America and southwestern Europe. These carnivores were typically found hunting in woodland areas, using their sharp, pointed teeth and strong jaws to swallow large chunks of food.

The hunting behavior of the Allosaurus is sneaky because they would hide in the plant undergrowth to surprise packs of herbivores moving through the area. This was necessary because their heavy weight kept them from moving quickly.

SIZE

Length: 40 Feet
Height: 18 Feet
Weight: 4,000 Pounds

NICKNAME

Other Lizard

GEOLOGICAL TIME

Late Jurassic Period

DISCOVERY

North America (1877)

UNIQUE FEATURES

Two short horns above and in front of each eye

DIMETRODON

di-MET-ro-DON

FUN FACT!

An adult could take up to 90 minutes to raise its body temperature 10 degrees.

How They Lived

Dimetrodon was found in swampy areas of what is now North America and Europe. They were cold-blooded creatures and were most active during the daytime.

An apex predator in its habitat, Dimetrodon was believed to be a very fast runner that preyed on fish, reptiles, amphibians, and insects. These creatures had two types of teeth for eating: sharp canines for digging into prey, and shearing teeth for grinding it up.

SIZE

Length: 12 Feet
Height: 4 Feet (at the hips)
Weight: 500 Pounds

NICKNAME

Two Measures Of Teeth

GEOLOGICAL TIME

Permian Period

DISCOVERY

North America (1878)

UNIQUE FEATURES

Walked with an ambling, splay-footed gait (like a crocodile)

NEXT GENERATION SCIENCE STANDARDS

The Next Generation Science Standards (NGSS) are a comprehensive set of standards that span K-12 science content. Every grade level and scientific discipline are covered and help to set expectations of what students should learn during their formative years. Our dinosaur AR book focuses on several elementary-age standards, and a couple of middle school ones, too. There are many opportunities to expand on these and have deeper discussions depending on the learner's age level and curiosity. The NGSS identified on these pages align to the various content sections provided within this text.

VELOCIRAPTOR

3-LS2-1. Construct an argument that some animals form groups that help members survive.
3-LS4-1. Analyze and interpret data from fossils to provide evidence of the organisms and the environments in which they lived long ago.
3-LS3-1. Analyze and interpret data to provide evidence that plants and animals have traits inherited from parents and that variation of these traits exists in a group of similar organisms.
3-LS4-2. Use evidence to construct an explanation for how the variations in characteristics among individuals of the same species may provide advantages in surviving, finding mates, and reproducing.
4-ESS1-1. Identify evidence from patterns in rock formations and fossils in rock layers to support an explanation for changes in a landscape over time.
4-ESS2-2. Analyze and interpret data from maps to describe patterns of Earth's features.
MS-LS2-2. Construct an explanation that predicts patterns of interactions among organisms across multiple ecosystems.
MS-LS4-2. Apply scientific ideas to construct an explanation for the anatomical similarities and differences among modern organisms and between modern and fossil organisms to infer evolutionary relationships.

www.ingramcontent.com/pod-product-compliance
Ingram Content Group UK Ltd.
Pitfield, Milton Keynes, MK11 3LW, UK
UKHW050142280726
14058UKWH00006B/787